COOKING WITH MATH!

MAKING DESSERTS WITH MATH!

By Santana Hunt

Gareth Stevens PUBLISHING

leveled reader math

Please visit our website, www.garethstevens.com. For a free color catalog of all our high-quality books, call toll free 1-800-542-2595 or fax 1-877-542-2596.

Library of Congress Cataloging-in-Publication Data
Names: Hunt, Santana, author.
Title: Making desserts with math! / Santana Hunt.
Description: New York : Gareth Stevens Publishing, [2020] | Series: Cooking with math! | Includes bibliographical references and index.
Identifiers: LCCN 2019018661| ISBN 9781538245538 (pbk.) | ISBN 9781538245552 (library bound) | ISBN 9781538245545 (6 pack)
Subjects: LCSH: Mathematics–Juvenile literature. | Cooking–Juvenile literature. | Desserts–Juvenile literature.
Classification: LCC QA40.5 .H86 2020 | DDC 510–dc23
LC record available at https://lccn.loc.gov/2019018661

Published in 2020 by
Gareth Stevens Publishing
111 East 14th Street, Suite 349
New York, NY 10003

Designer: Katelyn E. Reynolds
Editor: Kate Mikoley

Photo credits: Cover, p. 1 ffolas/Shutterstock.com; pp. 1–24 (gingham background) Mika Besfamilnaya/Shutterstock.com; pp. 1–24 (recipe background) A. Zhuravleva/Shutterstock.com; p. 5 Monkey Business Images/Shutterstock.com; p. 7 Brent Hofacker/Shutterstock.com; p. 8 Danielle Balderas/Shutterstock.com; p. 11 Vezzani Photography/Shutterstock.com; p. 15 Rimma Bondarenko/Shutterstock.com; p. 16 RTimages/Shutterstock.com; p. 19 iuliia_n/Shutterstock.com.

Printed in the United States of America

CPSIA compliance information: Batch #CW20GS: For further information contact Gareth Stevens, New York, New York at 1-800-542-2595.

CONTENTS

Yummy Math! 4

No-Bake Fudge 6

Peanut Butter Kiss Cookies 10

Cake Bars 14

Yogurt-Dipped Bananas 18

Glossary 22

Answer Key 22

For More Information 23

Index 24

Boldface words appear in the glossary.

Yummy Math!

In order to follow a great **recipe** for a **dessert**, you need math. Measuring, telling time, subtracting, and multiplying are just some of the skills you'll use when making the treats. Who knew math class could produce such tasty results?

Ask an adult before using the oven!

No-Bake Fudge

Measuring is an important part of making desserts. You will use measuring cups and spoons. Common units, or the amounts used for measuring, used when making desserts are cups, tablespoons, and teaspoons. Some recipes may call for a scale to weigh **ingredients**!

Your measuring spoon measures 1 teaspoon. Your measuring cup measures 1 cup. This recipe calls for 2 cups of chocolate chips. Which measuring tool will you use? How many times will you need to fill it? Check your answers on page 22!

NO-BAKE FUDGE

Ingredients:

14-ounce can of **sweetened condensed milk**
2 cups chocolate chips
1 teaspoon vanilla **extract**

Instructions:

1. Use butter or cooking spray all over a 9 x 9-inch (23 x 23 cm) cake pan.
2. Put chocolate chips and milk into a large microwave-safe bowl. Microwave it for about 1 minute. Stir until smooth and mixed well. Heat for another 30 seconds if needed.
3. Add vanilla extract.
4. Pour mixture into container. Allow to cool before cutting.

Peanut Butter Kiss Cookies

A recipe may call for a certain number of an ingredient. You may have more than you need. You'll need to use subtraction to find out the right amount. Save the leftovers for another recipe!

This recipe calls for 30 chocolate kisses. If you roll the mixture of ingredients, called dough, into bigger cookies, you may only need 20 kisses. How many kisses will you have left over?

30 chocolate kisses
- 20 chocolate kisses

? chocolate kisses

PEANUT BUTTER KISS COOKIES

Ingredients:

1 cup peanut butter
1 cup sugar
1 large egg
1 teaspoon vanilla extract
30 chocolate Kisses

Instructions:

1. Have an adult heat the oven to 350°F (177°C). Mix peanut butter, sugar, egg, and vanilla in a bowl. Roll dough into small balls and place on a baking sheet.
2. Bake cookies 10 to 12 minutes. Unwrap chocolates while you wait.
3. Push one chocolate Kiss into each baked cookie. Let cool.

Cake Bars

Recipes commonly give a **range** of time for how long a baked treat will be in the oven. You may use a timer to let you know when to check your cookies or cake. Still, you need to keep an eye on the clock!

This recipe says the cake bars will be done baking after 18 to 20 minutes. This clock shows the time you need to take the cake bars out of the oven.

What time does it show?

CAKE BARS

Ingredients:

1 box yellow cake mix
1 large egg
1 cup white chocolate baking chips
1/2 cup milk
1/3 cup canola oil
1/3 cup rainbow sprinkles

Instructions:

1. Have an adult heat the oven to 350°F (177°C). Mix together all ingredients, except the sprinkles.

2. Spray a 15 x 10 x 1-inch (38 x 25 x 2.5 cm) baking pan with cooking spray. Pour in the mixture and spread evenly.

3. Bake 18 to 20 minutes. Put sprinkles on top. After it cools, cut into bars.

Yogurt-Dipped Bananas

Many recipes tell you how many treats you can expect to have when you're done. You may want to make more! You can do this by doubling the recipe, or using two times the amount of each ingredient. You can use multiplication to figure it out!

In this recipe, you'll cut bananas into two pieces. The recipe makes eight frozen treats! How many treats would you get if you doubled the number of bananas?

$$\begin{array}{r} 8 \text{ bananas} \\ \times\ 2 \text{ pieces} \\ \hline ?\ \text{frozen treats} \end{array}$$

YOGURT-DIPPED BANANAS

(makes 8 treats)

Ingredients:

4 bananas, peeled
6-ounce cup of yogurt, any flavor
2 cups **cereal** or chopped nuts
8 wood craft sticks

Instructions:

1. Spoon yogurt and cereal or nuts into separate bowls. Line a baking sheet with wax paper.
2. Cut bananas in half and put one craft stick into the bottom of each.
3. Dip bananas in yogurt. Then, roll each banana in cereal or nuts.
4. Place coated bananas on baking sheet. **Freeze** for about an hour.

Glossary

cereal: a breakfast food made with grain

dessert: sweet food often eaten after a main meal

extract: matter pulled from something. In cooking, often a flavor.

freeze: to put something in a very cold place, such as a freezer, to make it solid

ingredient: a food that is mixed with other foods

range: a series of numbers that includes the highest and lowest amounts possible

recipe: an explanation of how to make food

sweetened condensed milk: canned milk with added sugar and a lot of the water taken out

Answer Key

p. 8 The measuring cup. You will fill it two times.

p. 12 10 chocolate kisses

p. 16 3:20 p.m.

p. 20 16 frozen treats

For More Information

Books

Eboch, M. M. *Crave-Worthy Candy Confections with a Side of Science*. North Mankato, MN: Capstone Press, 2019.

Levit, Joseph. *Let's Explore Math*. Minneapolis, MN: Lerner Publications, 2019.

Websites

2nd Grade Math
www.mathplayground.com/grade_2_games.html
Find math games here!

Recipes & Cooking for Kids
kidshealth.org/en/kids/recipes/
Discover more recipes you can make on this website.

Publisher's note to educators and parents: Our editors have carefully reviewed these websites to ensure that they are suitable for students. Many websites change frequently, however, and we cannot guarantee that a site's future contents will continue to meet our high standards of quality and educational value. Be advised that students should be closely supervised whenever they access the internet.

Index

bananas 20, 21

cake bars 14, 16, 17

clock 14, 16

cookies 10, 12, 13, 14

cup 6, 8, 9, 13, 17, 21, 22

dough 12

fudge 6, 9

leftovers 10

measuring 4, 6, 8

multiplication 18

oven 5, 13, 14, 16, 17

recipe 6, 8, 10, 12, 14, 16, 18, 20

scale 6

subtraction 10

teaspoon 6, 8, 9, 13

time 4, 8, 14, 16, 18